THE POLLUTION PROBLEM

HARMING OUR HABITATS

Enslow PUBLISHING

BY ANNA MCDOUGAL

Please visit our website, www.enslow.com. For a free color catalog of all our high-quality books, call toll free 1-800-398-2504 or fax 1-877-980-4454.

Library of Congress Cataloging-in-Publication Data
Names: McDougal, Anna, author.
Title: The pollution problem / Anna McDougal.
Description: Buffalo, NY : Enslow Publishing, [2026] | Series: Harming our habitats | Includes bibliographical references and index.
Identifiers: LCCN 2024039915 (print) | LCCN 2024039916 (ebook) | ISBN 9781978543331 (library binding) | ISBN 9781978543324 (paperback) | ISBN 9781978543348 (ebook)
Subjects: LCSH: Pollution–Juvenile literature. | Pollution prevention–Juvenile literature. | Refuse and refuse disposal–Juvenile literature.
Classification: LCC TD176 .M33 2026 (print) | LCC TD176 (ebook) | DDC 363.73–dc23/eng/20240923
LC record available at https://lccn.loc.gov/2024039915
LC ebook record available at https://lccn.loc.gov/2024039916

Published in 2026 by
Enslow Publishing
2544 Clinton Street
Buffalo, NY 14224

Portions of this work were originally authored by Barbara Linde and published as *Pollution*. All new material in this edition is authored by Anna McDougal.

Designer: Leslie Taylor
Editor: Caitlin McAneney

Photo credits: Cover, p. 1 (photo) Dennis MacDonald/Shutterstock.com; Series Art (HOH logo) AI Generated art/Shutterstock.com; Series Art (peach shape) chyworks/Shutterstock.com; Series Art interior (planet earth) Y Salnikova/Shutterstock.com; Series Art (background earth/sky) Background Land/Shutterstock.com; p. 5 Julian Peters Photography/Shutterstock.com; p. 7 MargJohnsonVA/Shutterstock.com; p. 8 Chris J Ferro/Shutterstock.com; p. 11 Wirestock Creators/Shutterstock.com; p. 13 Halfpoint/Shutterstock.com; p. 15 Drawii/Shutterstock.com; p. 17 Marek Masik/Shutterstock.com; p. 19 Mathias Berlin/Shutterstock.com; p. 21 Ermolaev Alexander/Shutterstock.com.

Printed in China

CPSIA compliance information: Batch #QSENS26. For further information contact Enslow Publishing at 1-800-398-2504.

CONTENTS

Words in the glossary appear in **bold** type the first time they are used in the text.

POLLUTION EVERYWHERE

Think of a beautiful **habitat** in nature. You might imagine the mountains, the rainforest, or the ocean. Many plants and animals call these natural spaces home. Unfortunately, many of these places suffer from pollution.

THINK ABOUT IT!

Have you seen pollution around your community? What does it look like?

This rusty barrel is full of oil, which leaks into the water and harms the habitat.

Pollution comes in many forms. Air pollution comes from harmful particles, or bits, in the air. Water pollution comes from chemicals dumped into waterways. Solid waste, or trash, can clutter the ground and hurt the animals that try to eat it. Pollution is a big problem!

HABITATS AT RISK

How does pollution hurt a habitat? Solid waste and chemicals can get into the soil, killing plants. Animals get their heads stuck in trash such as plastic bags and bottles. They can choke on straws and bottle caps. Oil can stick to the wings of seabirds, making them unable to fly.

THINK ABOUT IT!

Imagine an oil spill in a coastal habitat. Oil spreads through the water and onto the shore. Could it hurt multiple habitats at once?

Some pollution is hard to see! Microplastics are bits of plastic so small that people and animals don't notice them. They're still very harmful.

In an ecosystem, plants and animals depend on one another. If one species, or kind, of plant or animal dies, other species suffer too. That can throw the whole habitat out of balance.

TRASH ALL AROUND

Some hikers live by the rule "leave no trace." That means anything they bring with them into nature, they take home with them.

Trash is a major problem in towns and cities. It's also a major problem in natural habitats. People throw things away when they're done with them. Those things end up in landfills, which may leak harmful **substances** into the earth. Other people throw things on the ground or in rivers.

If you go to a park, notice the trash you find. There might be bottles in the woods, diapers by the playground, and old fishing lines by the creek. This trash makes a big difference in a habitat.

THINK ABOUT IT!

If you can't find a trash can, you should still keep your trash off the ground. Why is it important to take your trash with you?

MOUNTAINS OF TRASH

Even the most **remote** places on Earth have pollution. The tallest peak on Earth is Mount Everest. This mountain is home to snow leopards and red pandas. It's also very polluted. Climbers leave used oxygen tanks and food wrappers behind as they climb. Some leave tents and clothes. It's so cold on the mountain that the **debris** doesn't break down.

THINK ABOUT IT!

How can hikers and climbers keep mountains free from trash?

Everest isn't the only messy mountain. Trash from hikers and campers is a problem all over the world. Wild animals may eat some of the trash. This can make them sick!

This photograph shows the trash on Mount Everest.

THE PLASTIC PROBLEM

Plastic is a big part of the pollution problem. Unlike other materials, such as wood, it takes hundreds of years to break down. And humans make more than 380 million tons of plastic year!

THINK ABOUT IT!

About 8.8 million tons of plastic is dumped in the ocean each year. How could this affect ocean creatures?

Many plastics are thrown into a landfill, left on the ground, or dumped into the water. Scientists think nearly half of Earth's **marine** mammals have eaten plastic or been trapped in it. A huge area of ocean called the Great Pacific Garbage Patch is littered with harmful plastics.

WATER POLLUTION

Water habitats include the ocean, wetlands, creeks, lakes, and rivers. They're home to many species of fish, mammals, birds, and amphibians. And every animal on Earth needs water to live.

THINK ABOUT IT!

Microplastics swirl around in water, harming the animals that live and eat there. How is this harmful to humans as well?

HOW MICROPLASTICS GET INTO FOOD

When microplastics get in the water, fish may eat them. Then the animals or people that eat the fish also eat the microplastics.

Water is polluted by chemicals and other **toxic** substances. These include **pesticides**, landfill leaks, and **sewage**. Oil may spill into the sea, which can kill animals such as sea otters.

AIR POLLUTION

Air pollution is a big problem in today's world. It harms animals in their habitats, as well as humans. In fact, about 7 million people die each year from issues related to air pollution.

Air pollution is caused by both solid and liquid particles in the air. These harmful particles come from factories, fires, and **exhaust** from cars and trucks. They make it hard to breathe. When certain toxic particles mix with water droplets in the air, they can fall as acid rain.

When toxic particles, smoke, and fog come together, it makes smog. Smog harms many habitats around cities, such as near Los Angeles, California, shown here.

THINK ABOUT IT!

Acid rain pollutes waterways and soil. How could this harm food sources for people and animals?

LIGHT AND NOISE POLLUTION

Some kinds of pollution are hard to notice. But they can be very harmful to animals in their habitats. Light pollution comes from the lights of homes and businesses. It can confuse birds and sea turtles as they find their way.

THINK ABOUT IT!

The U.S. Navy uses sonar to find enemy submarines. Do you think it is worth it if sonar harms marine animals? Why or why not?

Whales and dolphins use sound to find food, find their way, and "talk" to each other. Noise pollution is very harmful to them. Some wash up on beaches and die.

Noise pollution is harmful too. Sonar and loud engines from motorboats, jet skis, and cruise ships cause terrible noise pollution problems. This harms animals that use sound to find food and get around.

POLLUTION SOLUTIONS

The pollution problem is a big one. No one person can fix it. However, we can all do something to help. You can try to use fewer new products, especially plastics. If you do use them, reuse them for something else or **recycle** them. Bring your own bags to the store. Carry a water bottle that can be reused.

You can also bike or walk instead of taking a car to help air pollution. Use less electricity by turning off lights in your house. You can make a difference!

Never throw trash on the ground or in the water. Pick up the things you see, then put them in the trash or recycling bin.

THINK ABOUT IT!

What kinds of pollution solutions can you try in your everyday life?

GLOSSARY

debris: The remains of something that has been broken.

exhaust: The gases that escape from the fuel burned in an engine.

habitat: The natural place where an animal or plant lives.

marine: Having to do with the ocean.

pesticide: Something used to kill pests, such as bugs.

recycle: To treat something so it can be used again instead of throwing it away.

remote: Far removed from populations of people.

sewage: Waste matter from buildings that is carried away through sewers.

sonar: A way of using sound waves to find things in a body of water.

substance: A certain kind of matter.

toxic: Being poisonous, especially when capable of causing death or harm.

FOR MORE INFORMATION

Books

Howell, Izzi. *Studying Pollution: Collect and Share Data*. New York, NY: PowerKids Press, 2023.

Phillips-Bartlett, Rebecca. *Stopping Pollution*. Minneapolis, MN: Bearport Publishing Company, 2024.

Websites

Light Pollution
kids.nationalgeographic.com/space/article/light-pollution
Discover a silent killer in habitats around the world: light pollution.

Pollution
kids.nationalgeographic.com/science/article/pollution
Learn more about pollution in habitats, especially the ocean.

Publisher's note to educators and parents: Our editors have carefully reviewed these websites to ensure that they are suitable for students. Many websites change frequently, however, and we cannot guarantee that a site's future contents will continue to meet our high standards of quality and educational value. Be advised that students should be closely supervised whenever they access the internet.

INDEX